OnBoard
ACADEMICS

Electricity and Magnets

© 2015 OnBoard Academics, Inc
Portsmouth, NH
800-596-3175
www.onboardacademics.com
ISBN: 978-1-63096-049-0

OnBoard Academic's books are specifically designed to be used as printed workbooks or as on-screen instruction. Each page offers focused exercises and students quickly master topics with enough proficiency to move on to the next level.

OnBoard Academic's lessons are used in over 25,000 classrooms to rave reviews. Our lessons are aligned to the most recent governmental standards and are updated from time to time as standards change. Correlation documents are located on our website. Our lessons are created, edited and evaluated by educators to ensure top quality and real life success.

Interactive lessons for digital whiteboards, mobile devices, and PCs are available at www.onboardacademics.com. These interactive lessons make great additions to our books.

You can always reach us at customerservice@onboardacademics.com.

Electric Circuits

www.onboardacademics.com

The parts of a simple electric circuit.

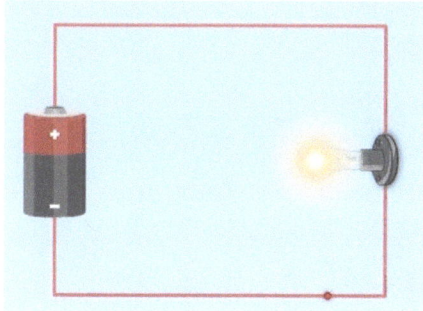

The elements that are required to make a bulb glow are: 1)the bulb itself and the appliance 2) a battery which is the power source 3) a wire which is the path to conduct the electricity.

When all of these three elements are in place we call this a circuit. A circuit enables the energy to flow from the minus or negative end of the battery to the plus or positive end of the battery.

If we want to turn the light bulb on and off we must add a switch to the circuit to interrupt the flow of electricity. When the switch is in the on position the circuit is complete and the energy can flow freely throughout the circuit making the bulb glow. We say that the circuit is closed.

When the switch is off the circuit is broken and electricity can not flow so the bulb does not glow. In this case, we say that the circuit is open.

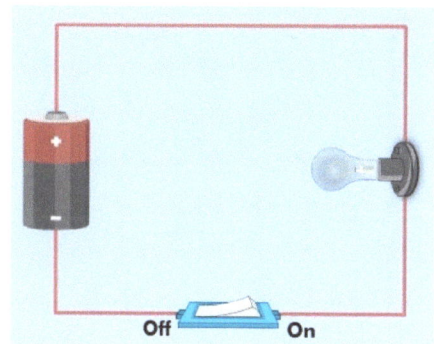

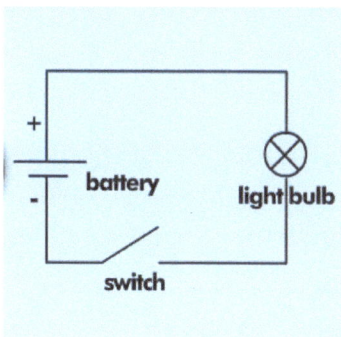

Electric circuits are represented using a circuit diagram. The battery is represented by a big line and a small line. The lightbulb is represented by a circle with an X inside of it. The switch is indicated by a line indicating an open or closed position.

Although electrons (tiny particles that carry an electric current) travel from the negative (-) terminal to the positive (+) terminal, the standard way to show the flow of an electric current is from positive to negative.

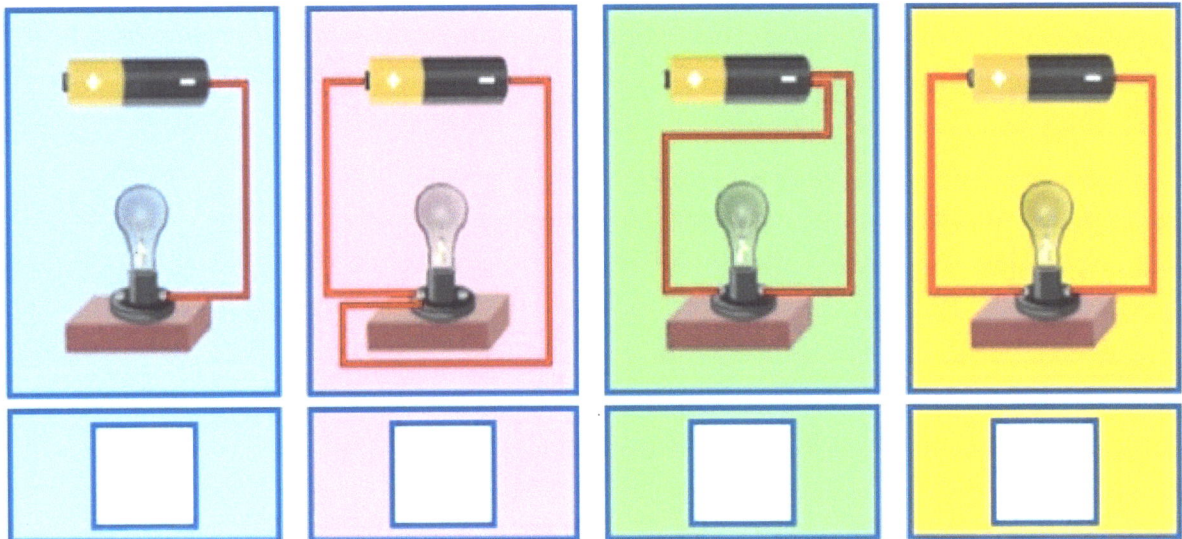

In order for electricity to pass through a circuit, the circuit must be complete. This means that electricity must be able to flow from the negative end of the battery, through the bulb and then back to the positive end of the battery.

Identify the parts of a circuit.

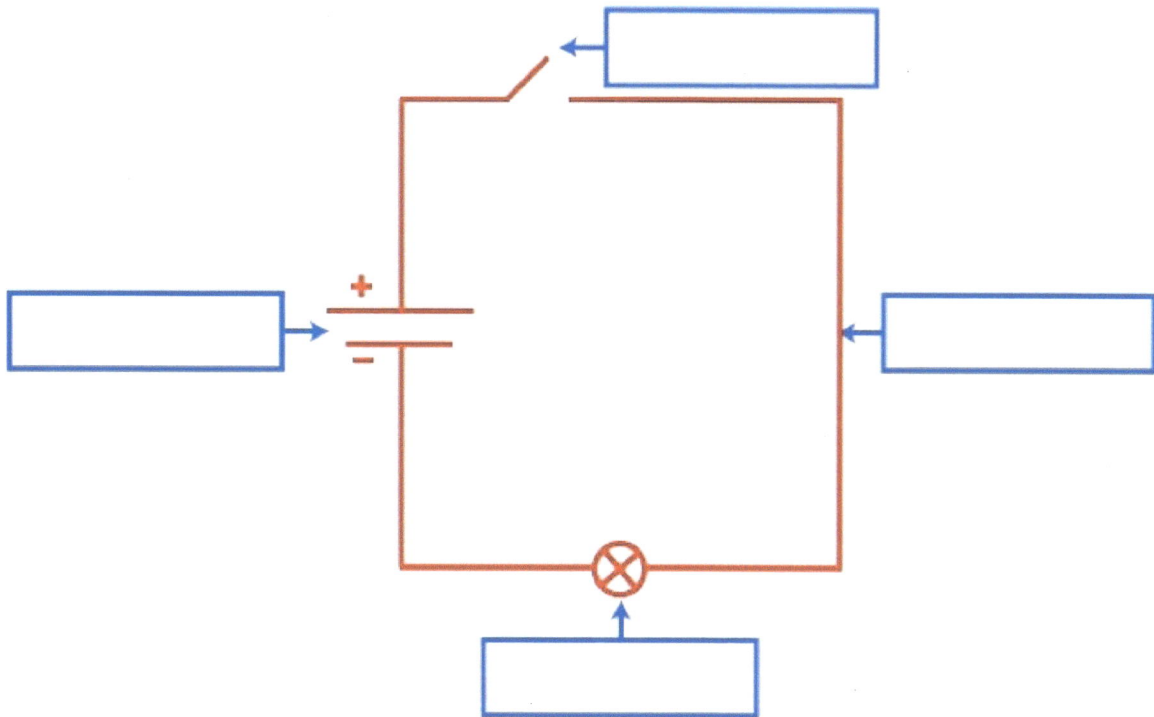

| Wire | Battery | Switch | Bulb |

The Effect of Changes to a Circuit

When you make changes to a circuit this can affect other elements of a circuit. For example if we were to add more wire, less electricity will reach the lightbulb causing a dimmer glow.

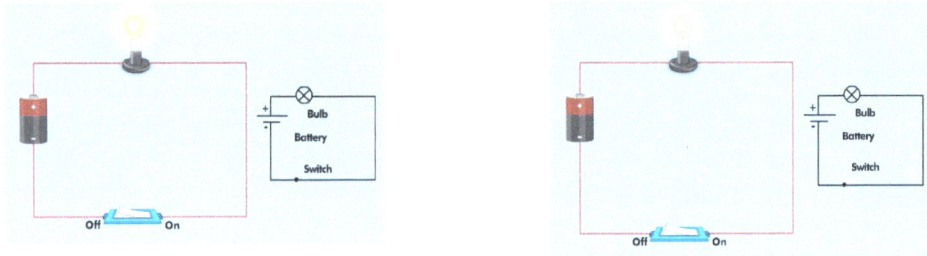

If we were to add another light bulb this too would reduce the intensity of the glow.

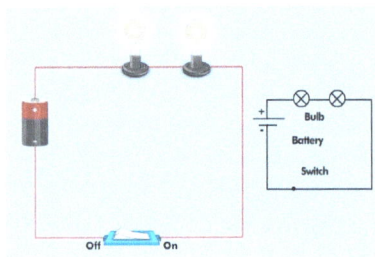

On the other hand, if we were to add another battery, this will increase the intensity of the lightbulbs' glow.

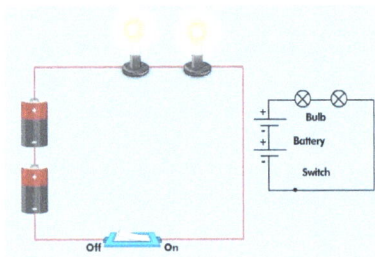

The effective changes to a circuit often depend on what type of circuit it is. Lets explore two types of circuits; series circuits and parallel circuits.

Series Circuits.

There are two main types of circuits; series and parallel.

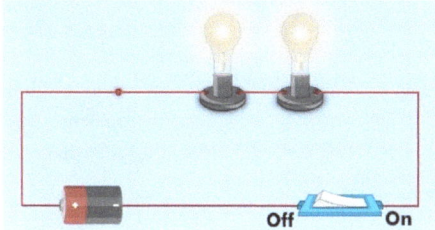

A series circuit is one in which light bulbs are added in a row so that electricity has to flow through one light bulb in order to get to another.

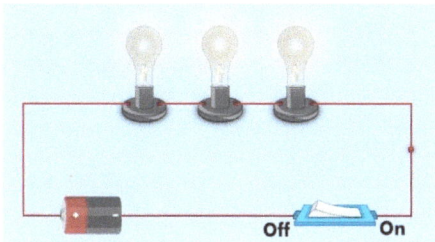

Since the electricity flows through one shared pathway the intensity of the glow decreases as more light bulbs are added to the circuit as electricity meets more resistance.

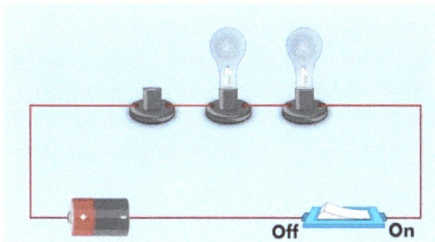

If a lightbulb in a series circuit breaks or is removed, all the lightbulbs will stop working because the circuit is broken and the electricity will not be able to flow in a complete loop.

In a series circuit, bulbs are added in a single row so that electricity has to flow through one bulb in order to get to the next bulb. Since the electricity flows along one shared pathway, the intensity of the glow decreases as more light bulbs are added, and the circuit is broken if a bulb breaks or is removed.

www.onboardacademics.com

Build a series circuit.

Draw the elements of the circuit onto the wire to represent a series circuit.

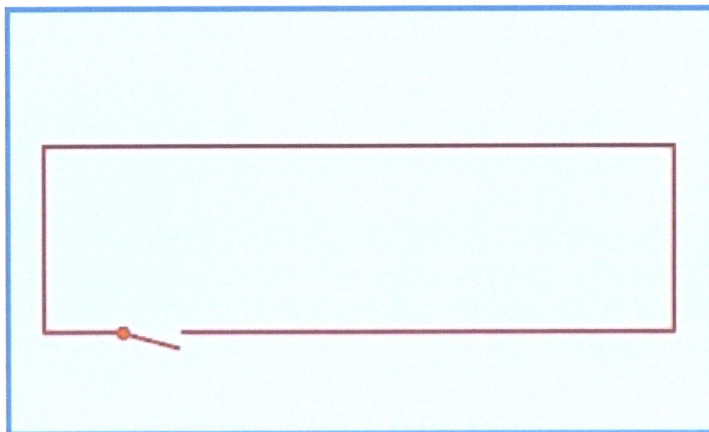

Battery

Bulb

Switch

The intensity of the glow decreases when more bulbs are added to the circuit. As you add batteries, the intensity of the glow increases equally for all the bulbs. However, too much electricity (such as two batteries for one bulb) could make a bulb burn out.

Parallel Circuits

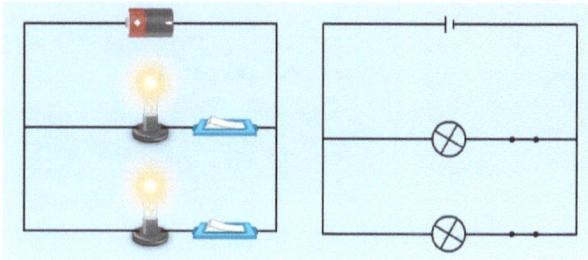

A parallel circuit has the same parts as a series circuit but it's constructed a little bit differently.

In a parallel circuit the electricity still flows from the negative to the positive end of the battery but the bulbs are added in separate parallel rows rather than in a single row.

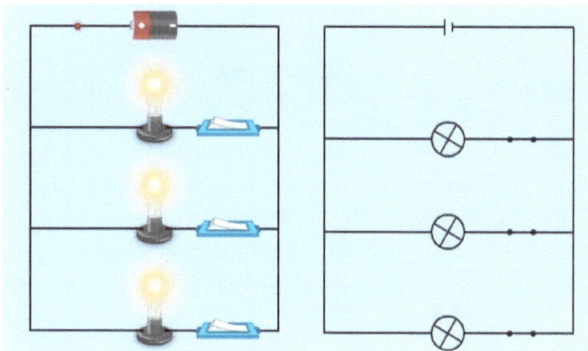

This means that each bulb has its own loop or circuit within the overall circuit and electricity flows separately through each of these mini circuits.

Because the electrical current that flows through each bulb is completely separate from the current that flows through the other bulbs the intensity of each bulb is not affected by adding other bulbs. This means that the bulbs intensity will not decrease by adding other bulbs.

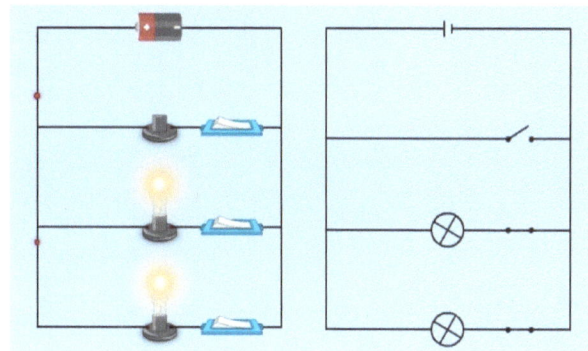

Also, if one bulb were to break or is removed, the other bulbs would continue to work since the other bulb's circuits remain closed.

It's for these reasons that parallel circuits are widely used.

> **In a parallel circuit, bulbs are added in separate parallel rows rather than in a single row. Because each bulb has its own loop, or circuit, the intensity of the glow of a bulb is not affected by adding more bulbs, and if a bulb breaks or is removed, the other bulbs will continue to work.**

Build a parallel circuit.

Draw the elements of a parallel circuit onto the wire diagram.

Battery

Bulb

The intensity of the glow is equally distributed across all the bulbs in a parallel circuit and as you add batteries, the intensity of the glow increases for all the bulbs equally. However, too much electricity per bulb (such as two batteries for one bulb) could make a bulb burn out.

www.onboardacademics.com

Series or Parallel Circuit?
Label each circuit fact with a P or an S.

○ As more bulbs are added to this circuit the light intensity of each bulb decreases.

○ If one bulb is removed from this circuit the other bulb continues to work.

○ Electricity flows in one single pathway in this circuit.

○ Lightbulbs can be added in separate pathways in this circuit.

○ If one bulb breaks in this circuit the other bulbs will not work.

○ As more bulbs are added to this circuit the light intensity of each stays the same.

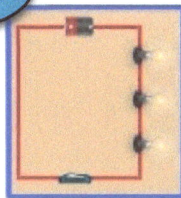

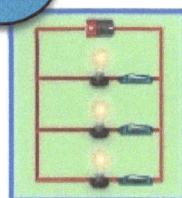

S   P

Electrical Circuits Quiz

1. Electric current is the flow of _____.
   - a. protons
   - b. neutrons
   - c. electrons
   - d. atoms

2. A _____ is a device used to open and close a circuit.
   - a. bulb
   - b. cell
   - c. switch
   - d. wire

3. As long as a batter and wire are connected in a circuit, current will flow even if there is no switch. True or false?

4. In a circuit, the flow of electrons is from the positive terminal to the negative terminal. True or false?

5. In a series circuit, electricity travels through _____.
   - a. one path
   - b. many paths
   - c. two different paths

Conductors and Insulators

Simple Circuit Parts Review

Label the parts of a simple circuit.

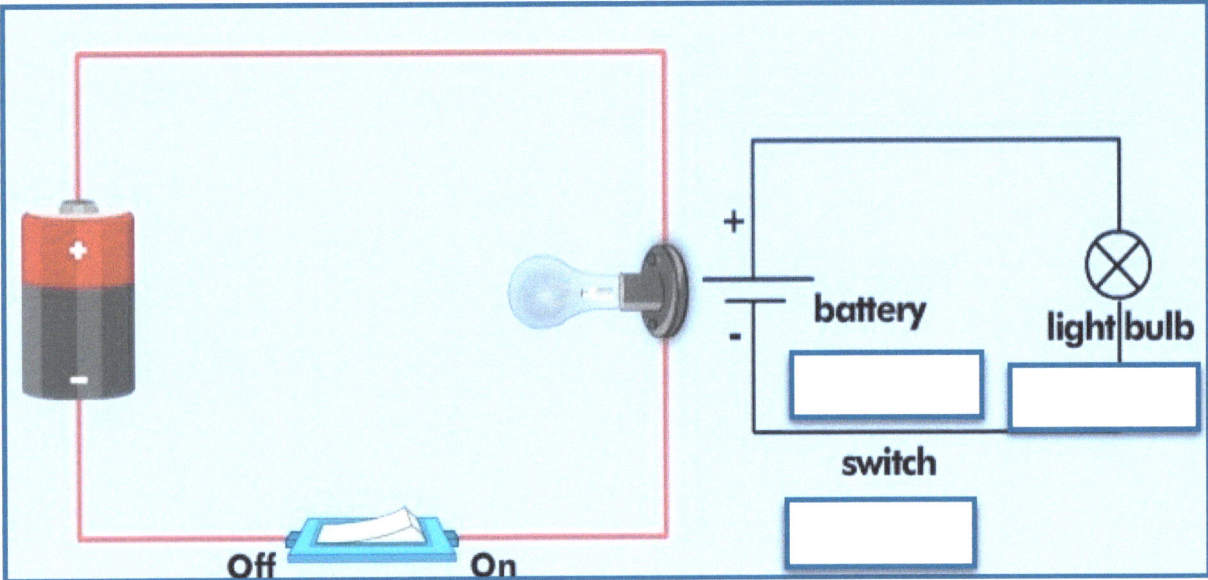

What are conductors and what are insulators?

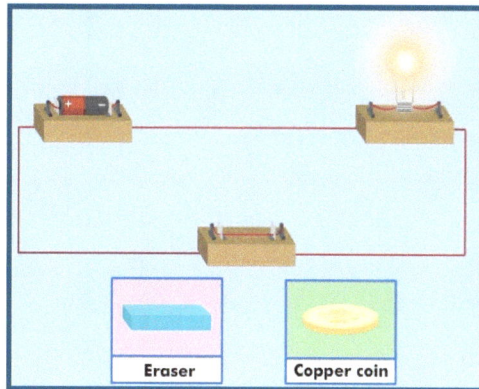

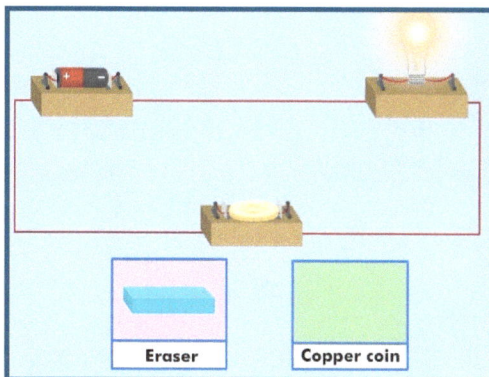

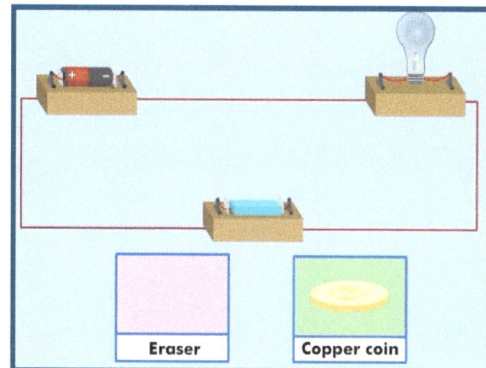

Look what happens when we add a copper coin or an eraser to a circuit.

Materials, like the copper coin, that allow an electric current to pass through them easily are called conductors. Materials, like the eraser, that don't conduct electricity well or at all are called insulators.

Sort the items by conductor or insulator.

conductor

insulator

gold
paper
copper
rubber
plastic
glass
aluminum
carbon
wood
styrofoam

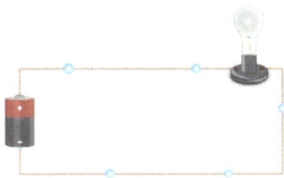

Why do conductors conduct electricity while insulators don't?

A chemical reaction in this battery creates tiny particles called electrons and these electrons travel from the minus side or negative end of the battery to the plus side or positive end of the battery. An electric current is the name we give to this energy that is carried through a conductor, a copper wire in this case, by these electrons.

To visualize how these electrons carry currents, lets imagine that the wire is a series of marbles. When the electron leaves the battery it bumps into the first marble that in turn bumps into the next marble and so on until the electron reaches the positive end of the battery to complete the circuit.

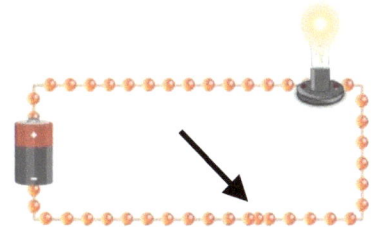

Materials such as copper, which are good conductors of electricity, has electrons that can move freely so can carry an electric current. Insulators such as wood hold on to their electrons tightly and therefore prevent the flow of energy.

Liquids and Electricity

We have taken a ci   onduct electricity.

pure water   salt water   water & lemon juice

pure water   **salt water**   **water & lemon juice**

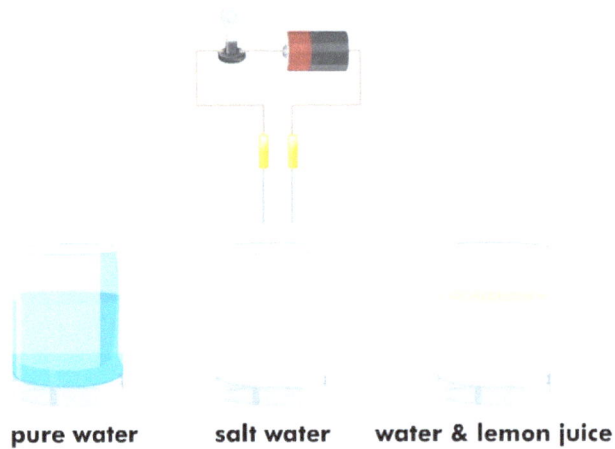

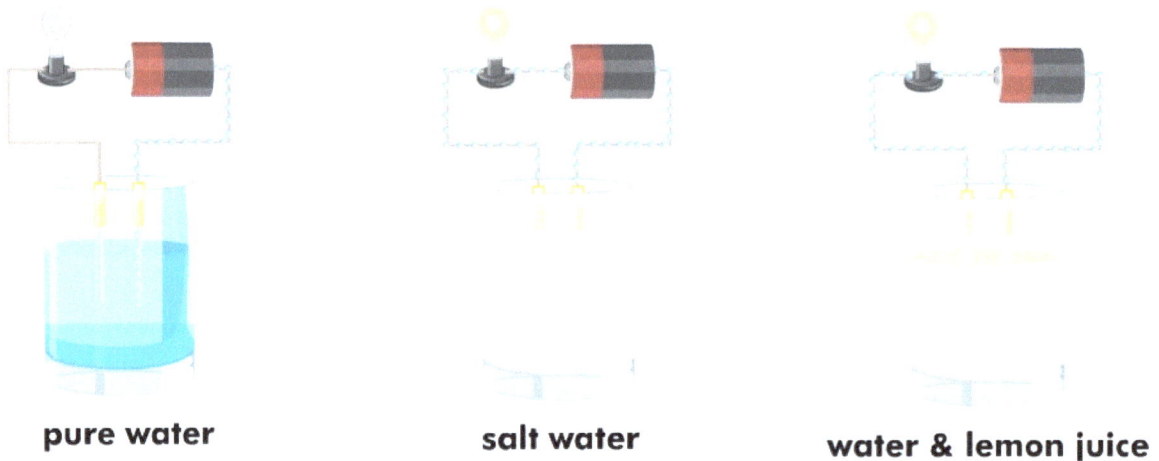

Pure water is not a good conductor of electricity. However, certain substances such as salts and acids break down in water to form charged particles which are able to conduct electricity through the solution. Since most water isn't pure, it's always important to be careful when using electrical devices near water.

Insulators or conductors of heat?

Feels Cool

Feels Warm

Feels Cool

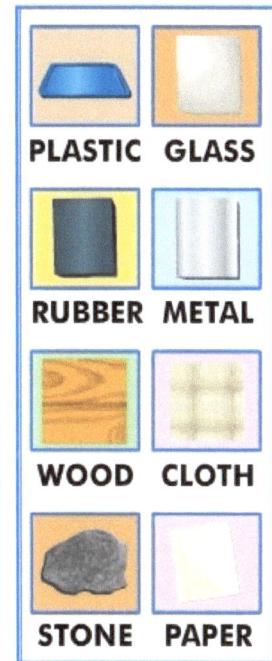

PLASTIC GLASS

RUBBER METAL

WOOD CLOTH

STONE PAPER

Feels Hot

Feels Cool

Feels Cool

Review how each of the materials conduct heat.

Feels Hot

Feels Cool

Most materials that are good conductors of electricity are also good conductors of heat, and most materials that are poor conductors of electricity are also poor conductors of heat. Glass is an insulator of electricity, but a weak conductor of heat.

Conductors and Insulators Review

Label each expressions as either "conductor" or "insulator."

An electric current can pass easily through these materials.	
This materials don't hold onto their electrons very tightly.	
These materials hold on tightly to their electrons.	
An electric current can't pass easily through these materials.	
Is the water in your bath a conductor or an insulator of electricity?	
Do good conductors of electricity tend to be good conductors or insulators of heat?	

conductor(s)

insulator(s)

 www.onboardacademics.com

Conductors and Insulators Quiz

1. The standard way to show the flow of an electric current is
   a. negative to positive
   b. positive to negative

2. The objects that allow electric current to pass through them easily are called _____.

3. Plastic and rubber are examples of conductors. True or false?

4. The energy that is carried through a conductor is called an electric current. True or false?

5. Electrons in wood can move freely. True or false?

6. _____ is (are) not a good conductor(s) of electricity.
   a. lemon juice
   b. salt water
   c. pure water
   d. electrolytes

Magnets

www.onboardacademics.com

James dropped a paper clip and a key into a drain. James had the bright idea to retrieve them using a magnet. James is a thinker! However, James retrieved the paper clip but the magnet was not attracted to the metal key.

Do you know why?_____

Magnets are objects that attract magnetic materials. The four most common magnetic materials are iron, nickel, cobalt, and steel (which is a mixture of iron and other materials). James can use the magnet to retrieve his paper clip, which is made of steel, but the key is made mostly of brass, which is not magnetic.

A magnet has two poles—one at each end—called a north pole and a south pole. Opposite, or unlike, poles attract (pull towards each other), while similar, or like poles, repel (push away from) each other.

www.onboardacademics.com

Will these poles attract or repel?

Place an **A** for attract.
Place an **R** for repel.

1	N S	S N	
2	N S	N S	
3	S N	S N	
4	S N	N S	

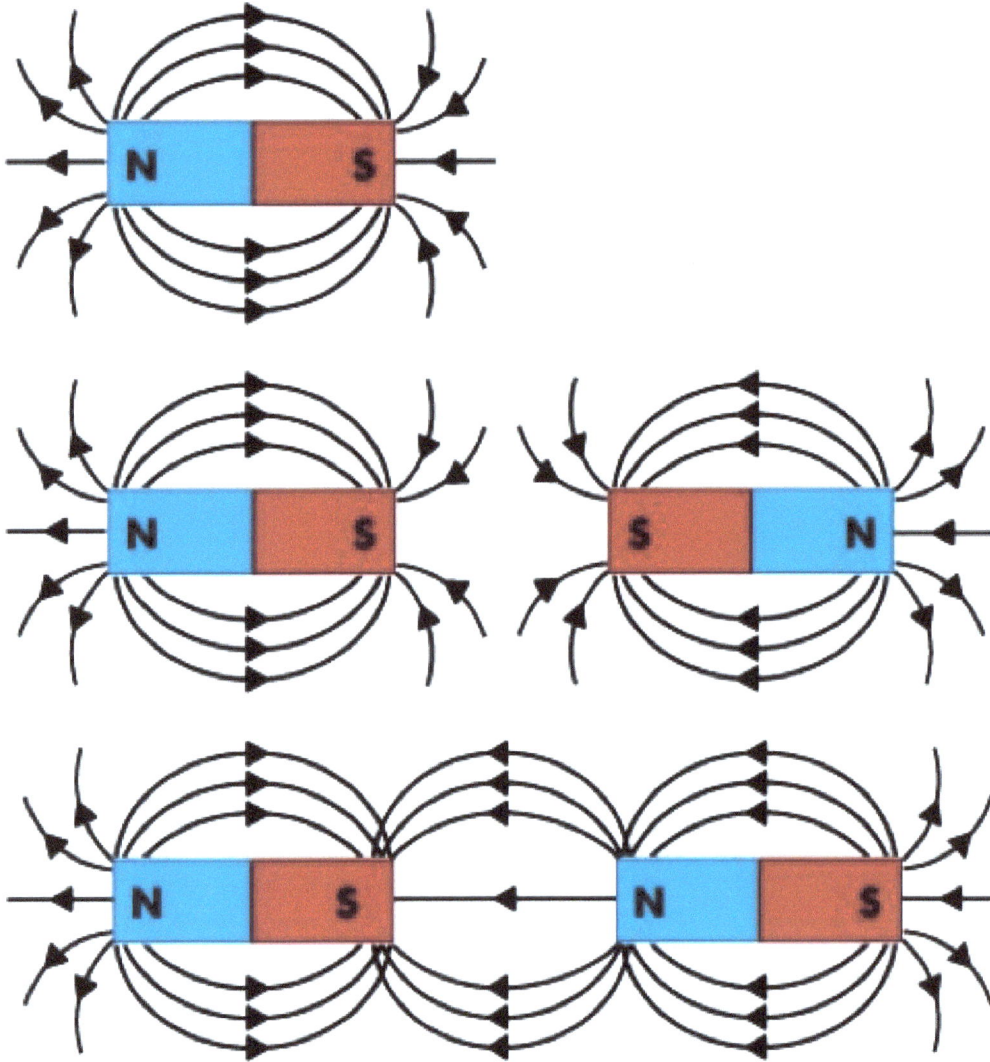

The area around a magnet where the force acts is called the magnetic field . The forces in a magnetic field can be demonstrated by sprinkling iron filings around a magnet. The filings show the forces of attraction and repulsion around a magnet which is strongest at the poles.

Circle the items that the magnet will pick up.

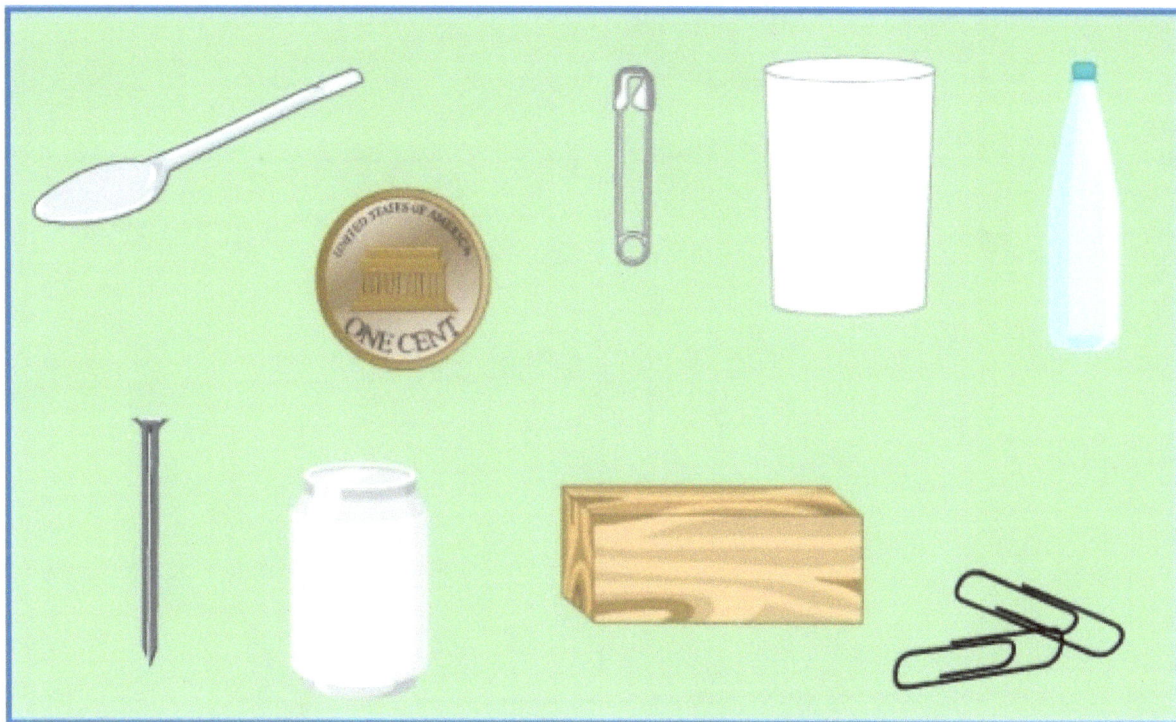

Metals such as iron, nickel, and cobalt all have
particles that react to a magnet's field and so we
call these metals magnetic. The particles in other
metals such as aluminum and copper don't react
to a magnet's field and so these metals aren't
magnetic. A penny is made up mostly of zinc
which isn't magnetic.

www.onboardacademics.com

Sort these materials by magnetic or non magnetic.

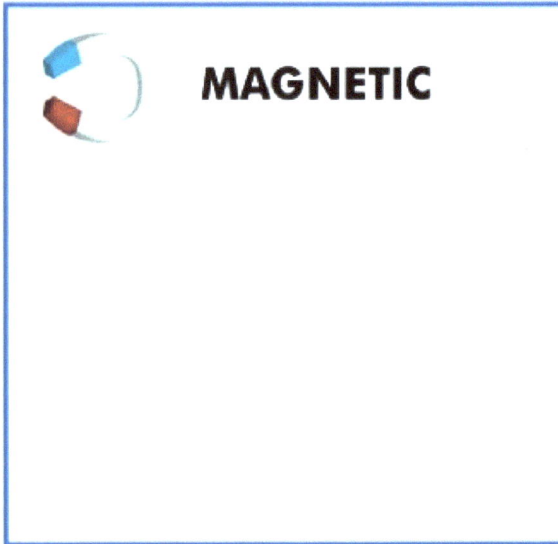

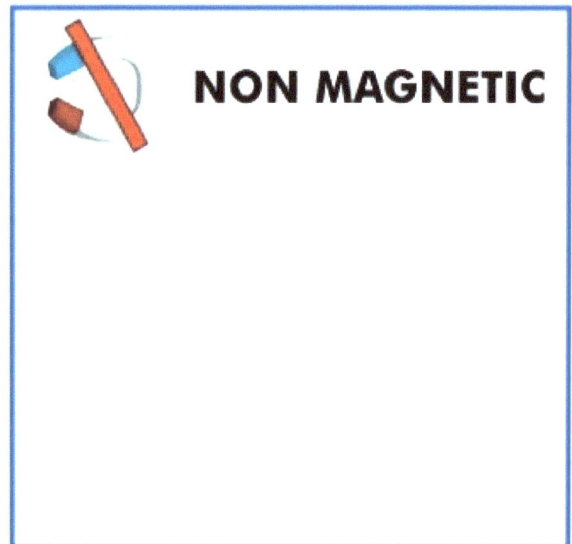

MAGNETIC	NON MAGNETIC

Steel   Silver   Rubber   Plastic

Gold   Aluminum   Cloth   Wood

Cobalt   Nickel   Paper   Iron

www.onboardacademics.com

The world's largest magnet.

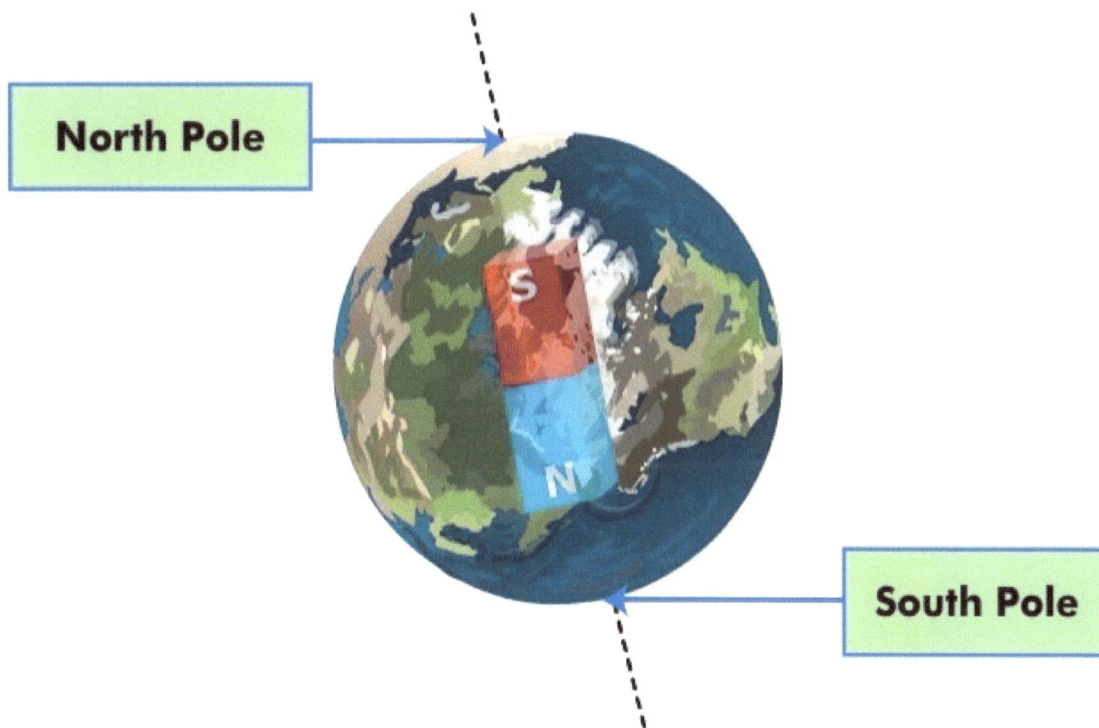

There is a large magnetic field surrounding the Earth. This is believed to be caused by the hot metals in Earth's core. As a result, the Earth itself acts like a magnet with two poles. A little confusingly, Earth's magnetic south pole is located somewhere close to Earth's geographic North Pole, while Earth's magnetic north pole is located close to Earth's geographic South Pole.

Compass

A compass uses a magnet that is shaped like a needle to give us directions. Because the end of the needle is actually the north-seeking pole of a magnet, it is always attracted to the Earth's magnetic south pole, which, as we've learned, is Earth's geographic North Pole. Once we know which direction north is, it's easy to identify south, west, and east.

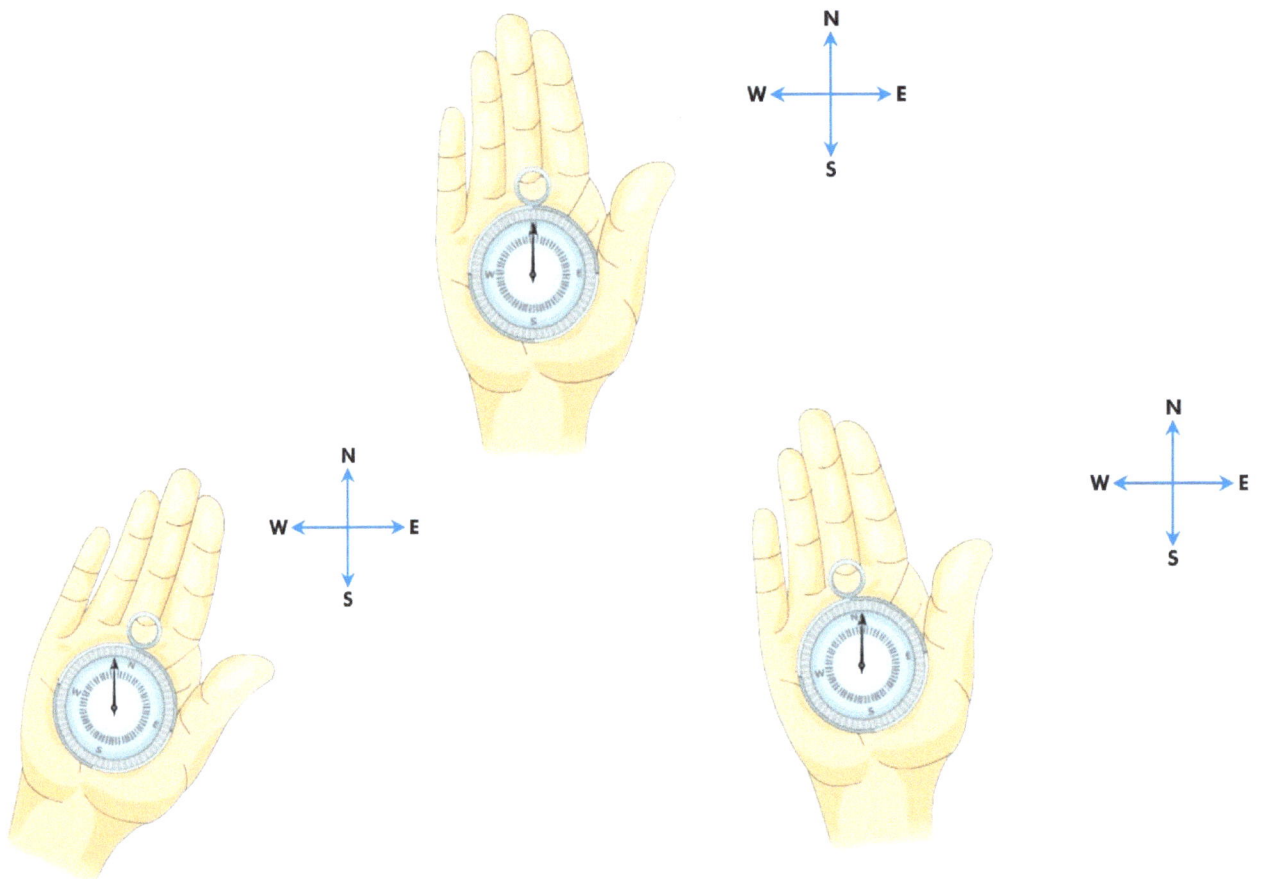

Apart from compasses where else are magnets found?

Place a √ for magnet
Place an X for no magnet

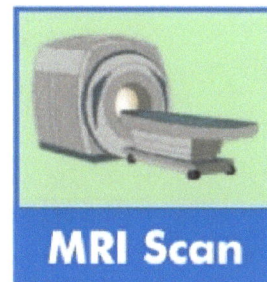

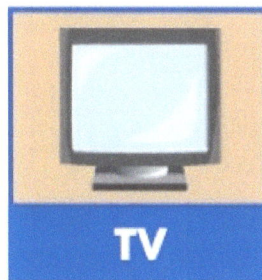

Dice	Maglev	Lamp	MRI Scan
☐	☐	☐	☐

Stove	TV	Stereo	ATM Card
☐	☐	☐	☐

Due to their ability to hold, control, and separate materials, magnets are used in many appliances including generators, telephones, televisions, and computers. Magnetism is also the force at work in Maglev trains, some of the fastest trains in the world.

www.onboardacademics.com

Magnets Quiz

1. Any magnet has two ends and they are called:
    a. magnetism
    b. magnetic material
    c. poles

2. The Earth's magnetic field is believed to be caused by the hot liquids in its core. True or false?

3. The like poles of two magnets attract each other. True or false?

4. The area around a magnet where the force of the magnetism can be experienced is called the _____.
    a. south pole
    b. north poe
    c. magnetic field
    d. magnetic personality

5. The magnetic force is stronger in the middle than at the poles. True or false?

6. A compass uses a magnet shaped like a needle to tell time. True or false?